THE
CLIMATE CHANGE
CONFLICT
Keeping Cool over Global Warming

JAKE HEBERT

INSTITUTE FOR
CREATION
RESEARCH

Dallas, Texas
ICR.org

Dr. Hebert earned a bachelor's degree in physics in 1995 from Lamar University and a master's degree in physics in 1999 from Texas A&M University. He received his Ph.D. in 2011 from the University of Texas at Dallas, where his research involved a cutting-edge study of the possible connection between cosmic rays, solar activity, and weather and climate. He joined the Institute for Creation Research in 2011 as a research associate, where he is helping to extend Dr. Larry Vardiman's work on climates before and after Noah's Flood, among other research endeavors.

THE CLIMATE CHANGE CONFLICT
Keeping Cool over Global Warming

by Jake Hebert, Ph.D.

First printing: March 2019

All Scripture quotations are from the New King James Version.

ISBN: 978-1-946246-28-8

Please visit our website for other books and resources: ICR.org

Printed in the United States of America.

TABLE OF CONTENTS

Introduction

Climate change is everywhere. Flip on the TV or click on your favorite online news site and you'll see headlines about how humans are drastically affecting Earth's climate, causing inestimable damage. Alarmists claim our planet is on the verge of near-destruction, and climate change skeptics are accused of being negligent—even dangerous. Climate change (or global warming) is definitely a "hot" topic, pardon the pun. But where did the idea of man-made climate change come from?

About 150 years ago, the Industrial Revolution took off. Humans began using fossil fuels such as oil, gasoline, and coal on a large scale. They powered steamships, factories, and many other things. Fossil fuels helped us explore unknown regions, increase trade between nations, and create new technologies.

But fossil fuels have a side effect. All that smoke coming from cars and factories contain *carbon dioxide*, a greenhouse gas. It traps heat and warms the earth, providing a greenhouse-like environment.

That's what everyone in the media is talking about. Some scientists think the fossil fuels we burn release so much carbon dioxide into the atmosphere that it's making Earth too hot. That's climate change.

Climate change is definitely a political issue, as we shall see. But is it scientific? It can be hard to navigate this hot topic. Politicians, the media, and many secular scientists say that climate change is a catastrophic issue and requires immediate action.

Christians have biblical reasons not to be concerned. After the catastrophe of the global Flood, Noah exited the Ark and made a sacrifice to God for His mercy. God responded with a rainbow—and an important promise. He promised He would never again destroy the Earth with a flood, adding, "While the earth remains, seedtime and harvest, cold and heat, winter and summer, and day and night shall not cease" (Genesis 8:22). This gives evidence that overall, Earth's climate is stable. God promised that Earth's seasons "shall not cease."

But the climate change issue can still be very confusing. How much of what we hear is just politics and how much is real science? Is the earth getting warmer? If so, is man-made carbon dioxide to blame? And can we do anything about it? *Should* we do anything about it? Culture and politics force everyone to have an opinion, even an uninformed one. Yet the answers often lie buried in technical data that can be understood only by a few highly technical individuals. This booklet covers all that and tries to make this tricky topic just a little more accessible.

1
Don't Panic!

In the spring of 2006, I was working as a physics lecturer and lab instructor at my *alma mater* Lamar University in Beaumont, Texas. One day, the university hosted a Texas Academy of Science meeting. I was not feeling well, so I went home as soon as I finished teaching. Soon after, I discovered an internet news story about a Dr. Eric Pianka of the University of Texas at Austin who made controversial statements about climate change. He said that Earth would be better off if 90% of humans were wiped off the face of the planet—perhaps by an airborne Ebola virus.

He got a standing ovation.

This was disturbing enough, but I was shocked to discover, from internet reports, that he made these very comments at my own Lamar University![1] The meeting I missed was the exact meeting where he made these controversial statements. Of course, I don't blame my former employer for Dr. Pianka's outrageous statements—Lamar University just happened to be hosting the Texas Academy of Science meeting that year.

The sentiments expressed by Pianka—whom some people call Dr. Doom—capture the hysteria of climate change: Humans are destroying planet Earth, and it would be better off without us. New York University bioethicist S. Matthew Liao suggested that we engineer humans to fight climate change. His ideas include altering genes to make people shorter, giving us drugs to take away our taste for meat, and giving us cat-like eyes so we can see in the dark to reduce our nighttime energy use and carbon footprint.[2] He wasn't joking.

From this radical perspective, climate change threatens every aspect of our lives in profound ways. Plenty of secular scientists think it's an important issue, and even evangelicals are raising questions about this topic. Many in Christendom express concern about climate change. Pope Francis urged action on the issue, and dozens of evangelical leaders signed a document stating that climate change is real and that Christians must take action.[3]

But is climate change real? Sort of. There are three things you should know about climate change:

1. The Earth has gotten warmer in the past, but this past warming was *not* due to human-produced carbon dioxide.

2. Carbon dioxide does cause some warming, but this warming is likely to be modest.

3. There are benefits to a warmer planet.

For much of the last century, the earth did get warmer. But this is not unusual. In fact, if you look at the temperatures of the Northern Hemisphere over the last 600 years, you'll see they fluctuated quite a bit (Figure 1).[4] Furthermore, between 950 and 1250 AD, Earth was warmer than it is today. Scientists call it the Medieval Warm Period. After that, temperatures dropped—a time known as the Little Ice Age. Then, around the turn of the 20th century, temperatures rose again. This means that Earth's temperature was fluctuating long before the Industrial Revolution, long before we humans could have been responsible for the warming.

This brings us to our second point. More carbon dioxide in the atmosphere *will* cause some warming, but that warming is likely to be modest. As climate experts acknowledge, the heart of the global warming debate is this: How sensitive is our climate to changes in carbon dioxide?[5] A gradual doubling of carbon dioxide in the atmosphere would cause only very modest warming, about 1°C (less than 2°F).[6] Obviously, this is nothing to panic over. In response, someone may point out that the climate is complex and that this estimation is too simple. However, as we'll see, even when we take the complexity of the climate into consideration, evidence can be found that carbon dioxide has very little effect on global warming.

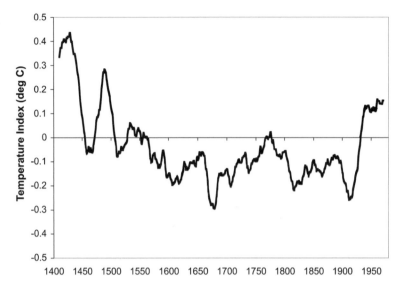

Figure 1. *Northern Hemisphere temperature variations in Celsius since 1400, showing that temperature swings are normal and no cause for alarm. Diagram adapted from McIntyre, S. and R. McKitrick. 2003. Corrections to the Mann. et al. 1998. Proxy Data Base and Northern Hemispheric Average Temperature Series.* Energy & Environment. *14 (6): 751-771.*

Scientists who are skeptical of the alarmism on this issue think Earth's climate is not as sensitive to changes in atmospheric carbon dioxide as many climate change advocates believe.[7] In other words, they think our climate is stable. In this booklet, I present evidence that this is indeed the case. I also explain why some disagree and why the arguments for an unstable climate are doubtful.

If Earth's climate is stable, then any future warming from increased carbon dioxide in the atmosphere will be modest. In fact, for over 20 years now, the warming has paused, and temperatures have actually decreased notably since 2016.[8,9]

But why do some scientists think our climate is *unstable*? There are three main reasons. The first involves the results from computer climate models that inaccurately predict dramatic warming from increased carbon dioxide. The second involves "junk science" of poor quality, including research that is flat-out fraudulent. The third is an old-earth interpretation of past climate data.

Some warming of the planet would actually be beneficial. Winter temperatures at some locations in Siberia occasionally drop below average temperatures on the surface of Mars.[10] The people of Siberia would no doubt welcome some heat! Warmer average temperatures would make more of the earth's surface habitable for people. Furthermore, far more people die prematurely from the cold than from the heat, so warmer temperatures would actually *save lives*.[11]

These are just brief answers to the question of climate change. Biblically, we would expect our climate to be stable because God has engineered marvelous design into our climate system. Earth's temperature levels have fluctuated throughout history, before we humans could have been responsible. Finally, there are benefits to a warmer climate.

Let's look at reasons why Earth's climate is stable.

2
Earth's Climate Is Stable

Climate sensitivity is the main issue in the debate over global warming. Alarmists think our climate is unstable and that increased carbon dioxide can cause a climate catastrophe. Those skeptical of the alarmism think Earth's climate is likely very stable and that the climate system self-adjusts to prevent extremes.

Climate sensitivity is defined as the average surface temperature increase after the amount of carbon dioxide in the atmosphere has been gradually doubled, and after the climate has had a chance to settle down from this change.[1]

American climatologist Judith Curry is skeptical of climate change alarmism. Dr. Curry has a stellar academic resume. She has authored nearly 200 papers, is the co-author or co-editor of two textbooks on atmospheric science,[2] and was awarded the 1992 Henry G. Houghton Research Award from the American Meteorological Society for promising young researchers.[3] Dr. Curry has provided expert testimony before congressional committees on the subject of climate change. Transcripts and videos of her testimony are posted online.[4,5]

She recently made a major contribution to the debate by presenting evidence that Earth's climate is stable, contrary to the claims of alarmists.[6] She released her results in a paper co-authored with Nicholas Lewis, an independent researcher who has also been very skeptical of climate change alarmism. The Lewis and Curry paper used historical climate data and a basic rule of physics called *energy balance* to estimate Earth's sensitivity to changes in carbon dioxide. Because their estimate was calculated from basic physics, it should be more reliable than estimates obtained from computer models. If the amount of atmospheric carbon dioxide doubled, Lewis and Curry estimated that the global average surface temperature would increase by 3.0°F (1.66°C). To put this number in perspective, most estimates of climate sensitivity estimate this temperature rise to be between 2.7°F (1.5°C) and 8.1°F (4.5°C). Alarmists on the issue typically argue for a climate sensitivity of about 5.4°F (3°C), and some claim that a sensitivity higher than 8.1°F (4.5°C) can't be ruled out.[7] Hence, Curry and Lewis estimate Earth's climate to be relatively stable.

Climatologist, meteorologist, and former NASA scientist Dr. Roy Spencer heralded their results. He noted that Lewis and Curry took great pains to address uncertainties, as well as possible objections to their calculations. Although he did not consider their result airtight (due to some factors that can't really be known), he considered their result very important, going so far as to call it "seminal."[8]

In addition to Curry's research, there are more reasons to think that our climate is stable. Rather than being harmful, carbon dioxide actually stimulates Earth's vegetation and helps stabilize the climate in the process. Humans breathe in oxygen and breathe out carbon dioxide. But what "breathes" carbon dioxide? Plants! As carbon dioxide increases, so does plant growth. Some deserts are shrinking as the amount of surrounding vegetation increases. This increase in plant growth seems to be a response to increased atmospheric carbon dioxide.[9] When carbon dioxide in the atmosphere increases, the numbers of plants increase too. These plants take in carbon dioxide, lowering the amount of carbon dioxide in the atmosphere, bringing it back toward its original values.

As noted by respected physicist Freeman Dyson:

> We know that plants do react very strongly to enhanced carbon dioxide. At [the Institute for Energy Analysis], they did lots of experiments with enhanced carbon dioxide and it has a drastic effect on plants because it is the main food source for the plants...So if you change the carbon dioxide drastically by a factor of two, the whole behavior of the plant is different. Anyway, that's so typical of the things [climate

alarmists] ignore. They are totally missing the biological side, which is probably more than half of the real system.[10]

Likewise, marine algae near the ocean surface trap carbon dioxide, and some scientists think that these algae help to regulate Earth's climate. Consider this scenario. As atmospheric carbon dioxide increases, temperatures warm. As temperatures warm, glacial ice melts. As this meltwater enters the ocean, so do iron-containing minerals. It is well known that iron in seawater increases the amount of algae. The iron stimulates "blooms" of algae, which use and trap the carbon dioxide, incorporating the carbon within their cells. When the algae die, they drift to the bottom of the ocean. This removes carbon dioxide from the atmosphere and sequesters it on the ocean floor. In describing this mechanism, one secular scientist said, "The Earth itself seems to want to save us [from global warming]."[11] Of course, it is God who should be given credit for the design of our climate system, not the earth!

It looks like carbon dioxide only has a modest effect on Earth's temperature. What else could be affecting our climate? The answer might lie 93 million miles away.

3

Could the Sun Be Affecting Weather and Climate?

Many scientists have long suspected that the sun could somehow be affecting Earth's weather and climate. Even the famous astronomer William Herschel dabbled in this area.[1]

One of the obvious difficulties with this idea is that the energy output of the sun changes very little over time. How then could the sun cause changes in weather and climate? One way is that the stream of charged particles from the sun (the solar wind) drags the sun's magnetic field so that it stretches throughout the solar system. The magnetic field at the earth's location does noticeably affect the number of cosmic rays (energetic charged particles from space) entering the earth's atmosphere. For instance, more cosmic rays enter the atmosphere during times of low solar activity (fewer sunspots) than when solar activity is high. For this reason, many suspect that the sun is indirectly affecting weather and climate by affecting the number of cosmic rays entering the atmosphere. These rays influence clouds, thus affecting weather and climate.

There are two main ideas about how the sun could influence clouds and consequently affect climate. One was popularized by Danish physicist Henrik Svensmark. In fact, ICR has previously discussed his work.[2] Svensmark thinks that cosmic rays stimulate the production of ions in the atmosphere and that these ions increase the number of clouds. But one of the major problems with Svensmark's hypothesis is that different datasets do not agree with his theory.

My Ph.D. advisor at the University of Texas at Dallas, Dr. Brian Tinsley, has a much more convincing explanation, even though it is less widely known. It explains a wide array of observations. (For detailed discussions of his hypothesis, see my 2013 and 2014 *Journal of Creation* articles listed at ICR.org/jake_hebert. To be clear, Dr. Tinsley does not share my creationist views.) Although Tinsley's idea is hard to explain, one of the ways it is obviously superior to Svensmark's is that the same predicted effect has shown up in multiple datasets. Even before I began working for him, Tinsley had ruled out other possible explanations for the sun/cosmic ray/weather connection, and had concluded that the real connection involved something called the *global electric circuit*. He reasoned that at certain times winter cyclones in the northern high latitudes would become more intense.

A big part of my research was to see if I could find evidence for this effect. The predicted effect was subtle and hard to see, but I eventually found evidence of the effect in two different datasets. These results, which flowed from my Ph.D. research, were published in 2012.[3,4] I even found preliminary evidence in a third dataset! Unlike Svensmark's, these results were *repeatable*, a strong indication that Tinsley was on the right track.

Unfortunately, the third result was arrived at quickly and never published in a paper. I found this result at about the time I was hired by ICR and had to abandon work on the subject. However, Dr. Tinsley and I presented all three results at a poster session of the 2011 American Geophysical Union meeting.[5]

Remember the Little Ice Age we mentioned earlier? European winters were quite cold during that period, much more so than in recent times. They were so cold that bodies of water froze, such as in Holland (Figure 2), even though those same bodies of water rarely freeze today. European winters were particularly cold during the part of the Little Ice Age called the Maunder Minimum.

The Maunder Minimum is named for the famous English astronomer Edward Maunder. It was a period lasting from 1645 to 1715 during which the sun had very few sunspots. Tinsley's hypothesis helps to explain the severity of European winters during times of low solar activity, including times of low sunspot number in recent years.

Figure 2. Winter Landscape with Skaters *by Dutch artist Hendrick Avercamp, 1608.*

During times when the sun has few sunspots, winter cyclones in the northern high latitudes become more intense. These cyclones are what meteorologists call *low pressure systems*. One of the results of this storm intensification is that high pressure systems intensify downstream from these storms and block the flow of warm moist air from the Atlantic onto Europe. This results in much more severe winters. Tinsley thinks this helps explain the severity of winters during times of low sunspot number, including the Maunder Minimum.

My research with Dr. Tinsley gave me a little firsthand experience in the way that politics influences climate research. When preparing my paper for publication, I suggested to Dr. Tinsley that we highlight the importance of this research in the paper's summary, or abstract. Clouds are one of the biggest uncertainties in computer climate models. Since our research involved clouds, I felt we should make that a selling point of the paper.

But Dr. Tinsley advised me not to do this! The reason was that he feared the reviewers might reject the paper if we stated this explicitly, regardless of the quality of the paper. Even though Tinsley believes in man-made global warming, his research implies that there are still things about weather and climate that scientists don't completely understand.

Some scientists are afraid to acknowledge this fact for fear it would give rhetorical ammunition to those who question alarmist claims.

So we have good reasons to think our climate system is stable, not unstable. But why do scientists insist that it is unstable? We'll see more on that in the next few chapters.

4

Faulty Computer Models

As stated in the first chapter, there are three main reasons scientists think Earth's climate is unstable. The first is the inaccurate results from computer climate models that predict dramatic warming due to increased carbon dioxide. The second involves "junk science." The third is an old-earth interpretation of past climate data.

Let's look at those computer models. Our climate system is so complicated that computer models must make simplifying assumptions, assumptions that could very well be in error. As noted by Dr. Roy Spencer:

> Remember, the sensitivity of their [computer climate] models is NOT the result of basic physics, as some folks claim… it's the result of very uncertain parameterizations (e.g. clouds) and assumptions (e.g. precipitation efficiency effects on the atmospheric water vapor profile and thus feedback). The models are adjusted to produce warming estimates that "look about right" to the modelers. Yes, *some* amount of warming from increasing CO_2 is reasonable from basic physics. But just *how much* warming is open to manipulation within the uncertain portions of the models.[1]

In fact, these assumptions are likely wrong. Most computer models have, in hindsight, greatly overestimated amounts of past warming. In 2013, the Intergovernmental Panel on Climate Change (IPCC) released a graph comparing global temperature measurements with various computer model predictions (Figure 3).[2] Note that when computer model

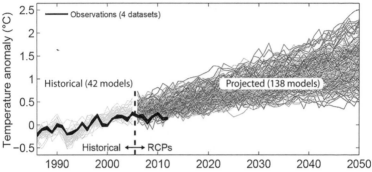

Figure 3. *ExxonMobil's compendium of observed data and model predictions, based upon IPCC reports, for global temperature change over the next 35 years.*

Image credit: Stocker, T.F. et al. 2013: Technical Summary. In *Climate Change 2013: The Physical Science Basis. Contribution of Working Group I to the Fifth Assessment Report of the Intergovernmental Panel on Climate Change.* Stocker, T.F. et al., eds. Cambridge University Press, Cambridge, United Kingdom and New York, NY, USA, pp. 33–115, doi:10.1017/CBO9781107415324.005.

results (thin lines) can be compared with actual observations (thick black line), the computer predictions were almost always higher than what was observed. This is an indication that the computer models are over-predicting the amount of future warming.

Dr. Spencer is absolutely right that clouds are one of the biggest uncertainties in climate models, and this is common knowledge among experts. Clouds can have either a warming or cooling effect depending on their altitude, latitude, and droplet sizes. Yet the physics of clouds, on a microscopic level, are poorly understood.

Respected physicist Freeman Dyson voices similar concerns:

> I have studied the climate models and I know what they can do. The models solve the equations of fluid dynamics, and they do a very good job of describing the fluid motions of the atmosphere and the oceans. They do a very poor job of describing the clouds, the dust, the chemistry and the biology of fields and farms and forests. They do not begin to de-

scribe the real world that we live in. The real world is muddy and messy and full of things that we do not yet understand. It is much easier for a scientist to sit in an air-conditioned building and run computer models, than to put on winter clothes and measure what is really happening outside in the swamps and the clouds. That is why the climate model experts end up believing their own models.[3]

The models are consistently unreliable, but many scientists still base their alarmism on them, causing a huge amount of hysteria.

Dr. Judith Curry, discussed previously, is a good example. Even though evidence for a stable climate exists, climate change has become so politicized that disagreeing with it is almost like committing a scientific crime. Dr. Curry has data to back up her skepticism, but she resigned from her tenured faculty position in frustration with the "craziness" of climate science. Dr. Curry stated in her blog:

> A deciding factor was that I no longer know what to say to students and postdocs regarding how to navigate the CRA-ZINESS in the field of climate science. Research and other professional activities are professionally rewarded only if they are channeled in certain directions approved by a politicized academic establishment.[4]

She is not alone. Other prominent scientists have also expressed concern that alarmists are being too dogmatic in their claims of the dangers of climate change. In 2011 physicist Ivar Giaever, a 1973 Nobel Prize winner, resigned as a Fellow of the American Physical Society in protest over its dogmatic stance on the issue.[5] Likewise, respected physicist Hal Lewis resigned from the American Physical Society for essentially the same reason.[6]

The resignation of several prominent scientists over the "craziness" of climate change is a clear indication of the politicization of this issue.

5

Politics and Junk Science

For decades, the Weather Channel has provided weather forecasts and weather-related news, and produced documentaries on weather and climate. It was founded by award-winning meteorologist John Coleman. As someone deep in media—*weather*-related media at that—you might think Coleman would have proclaimed the dangers of climate change. But he didn't. He remained a skeptic. When he retired, he claimed that the idea that humans are changing the climate with carbon dioxide is "nothing but a lie."[1] How could someone with his position and expertise be so skeptical? Some of it may concern the second reason why scientists think Earth's climate is unstable—poor-quality science and fraudulent research.

The influence of politics in the climate change debate is undeniable. But how much is politics and how much is science? Let's consider an interesting disagreement. In 2013, the IPCC released a report that accused man-made carbon dioxide emissions of being the "very likely" cause of global warming. They predicted, "Continued emissions of greenhouse gases will cause further warming."[2]

Another organization called the Nongovernmental International Panel on Climate Change (NIPCC) also released a report on climate change in 2013. In it, they pointed out that Earth's temperature leveled off in 1997, but carbon dioxide continued to increase by 8%. They said, "No close correlation exists between temperature variation over the past 150 years and human related CO_2 emissions."[3] It's interesting that a governmental organization concludes that man-made carbon dioxide causes global warming, but a *nongovernmental* organization concludes that it doesn't. Two organizations, two different conclusions. It looks like politics has plenty to do with climate change.

Unfortunately, many frauds have been discovered within the climate change alarmism community. A famous one concerns the well-known "hockey stick" diagram produced by Michael Mann in 1998. This diagram supposedly showed relatively uniform Northern Hemisphere temperatures from about 1000 to 1850 AD (the handle of the stick), then a sudden increase (the blade) following the Industrial Revolution. This diagram was touted as dramatic evidence of catastrophic global warming due to man-made carbon dioxide.

Although initially lauded, the hockey-stick diagram eventually came under intense scrutiny from other scientists. First, the "handle" part contradicts mainstream climate reconstructions that show large temperature fluctuations over that 1,000-year time period (Figure 4). The Medieval

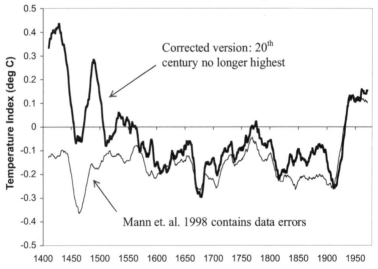

Figure 4. *Mann's "hockey stick" graph compared to McIntyre, S. and R. McKitrick results showing Earth's temperature over the last 600 years.*

Warm Period lasted from about 950 to 1250 AD. Crops and economies flourished, and Vikings explored and settled where ice sheets had melted. Things got colder during the Little Ice Age, which lasted from about 1300 to 1850 AD. During the Little Ice Age, rivers and canals in Europe routinely froze over, even though this rarely, if ever, happens today. It was especially cold between 1645 and 1715 AD, when the sun had very few sunspots. Yet Mann's graph made the Medieval Warm Period disappear!

Michael Mann's hockey-stick diagram contradicted widely known data so much that some secular researchers suspected fraud.[4] Two secular researchers were able to retrace Mann's steps so that they could very accurately reproduce his results. When they fed random sets of data into a code using Mann's method, it consistently produced hockey-stick graphs! The program suppressed data that did not produce a hockey stick and emphasized those that did. Mann's methodology was clearly in error.[5] In fact, some respected climate researchers have denounced him in blistering terms.[6] (As a side note, the IPCC endorsed the hockey-stick diagram.)

Then there's the notorious "climategate" email scandal. The University of East Anglia in England hosts a team of researchers called the Climatic Research Unit to study man-made climate change. In 2009, over a thousand sensitive emails and documents from the group were published online. Some contained dialogue exposing apparent deception by the researchers. The Director of the Unit, Phil Jones, sent emails that use words like "trick" and "hide" when discussing the presentation of climate data. One email states, "I've just completed Mike's Nature trick of adding in the real temps to each series for the last 20 years (ie from 1981 onwards) amd [*sic*] from 1961 for Keith's to hide the decline...."[7] Likewise, East Anglia scientists discussed using hardball tactics to suppress research that didn't support global warming.[8]

The University of East Anglia receives £35.6 million annually in governmental research grants.[9] That is equivalent to roughly 47 million U.S. dollars. Climate alarmists routinely charge that skeptics of global warming are in the pocket of "big oil," but it seems that climate alarmism can be rather profitable!

Maybe the failure of having supporting hard evidence is why climate change alarmists are taking the issue beyond science and into the courtroom. Twenty academics wrote former President Barack Obama, asking him to use the Racketeer Influenced and Corrupt Organizations (RICO) law to investigate climate change skeptics.[10] The letter urged "investigation of corporations and other organizations that have knowingly deceived the American people about the risks of climate change, as a means to forestall America's response to climate change."[11] Regarding climate change, a British professor of international law argued, "One of the most important things an international court could do—in my view it is probably the single most important thing it could do—is to settle the scientific dispute."[12]

The prosecution of climate change skeptics through court systems seems an act of desperation. Perhaps alarmists noticed that the data didn't support their position and that their predictions of increasing global temperature kept failing, and so they quietly went from using the term "global warming" to the more generic term "climate change." That term is far more convenient for them. Whether the Earth gets warmer or cooler, they can just call it "climate change."

6
Secular Climate Change Theories

The third and final reason scientists believe Earth's climate is unstable comes from their evolutionary worldview. This impacts climate change thinking through something called the Milankovitch (or astronomical) theory, which attempts to explain past ice ages by changes in sunlight from Earth's orbital and rotational motions. However, these changes are so ineffective at influencing Earth's climate that secular scientists conclude something else must be blamed. The journal *Science* writes:

> Paleoclimatologists [scientists who study ancient climates] have long recognized that the amount of Milankovitch-induced change in solar heating is too small to melt glaciers or to send Earth into a deep freeze, unless some as yet unidentified part of the climate system amplifies it.[1]

To them, human-produced carbon dioxide seems a likely culprit. This is responsible for much of the "craziness"—as Judith Curry calls it—of climate change. The next chapter will dig deeper into the connection between climate change alarmism and the Milankovitch theory, but first let's look at the theory itself and how it fails to explain ice ages.

We can define an ice age as a time when the earth's surface is covered by significant amounts of ice. There is strong geological evidence that about 30% was covered during the Ice Age.

Despite popular perception, secular scientists have a very difficult time understanding how an ice age happens. They believe Earth has experienced multiple ice ages. (Creation scientists think they are mistaken and are misinterpreting the evidence.) They say there have been at least five major ice ages, four of which supposedly occurred hundreds of millions of years ago. The most recent ice age is thought to have started 2.6 million years ago, at the beginning of what secular scientists call the Pleistocene Epoch. Since large ice sheets still cover about 10% of Earth's land surface, this ice age has not yet ended.

However, secular scientists believe that the ice sheets periodically grew and then shrank around 50 times during this most recent ice age. Scientists call the times at which ice sheets were larger *glacials*, or *glacial intervals*. These glacial intervals are what most people call "ice ages." Secular scientists say we are now living in a warm period with smaller ice sheets, which they call an *interglacial*.

Although secular scientists believe in multiple ice ages, explaining even a single ice age is very difficult for them. One might naively think all that is necessary for an ice age is very cold winters, but that isn't the case. For an ice age to occur, winter snow and ice must be prevented from melting during the summer months, and this must remain true for many years. If this happens, snow and ice will accumulate over successive winters, allowing the formation of thick ice sheets. Obviously, cooler summers are needed to make this happen.

One also needs heavy snowfall, because light snowfall will still tend to melt even during cool summers. But heavy snowfall requires mild—not cold—winters. This is because the moisture content of cold air is dramatically less than that of warmer air. Less moisture in the air means less precipitation, including less snowfall. So colder winters are actually a hindrance to an ice age, not a help!

And therein lies the problem. An ice age requires cold summers and moderate winters. However, one generally expects cold winters to go along with cool summers and warm summers to go along with warmer (or more moderate) winters, especially over a period of many years. Under normal conditions, one does not expect seasonal temperatures to continually "see saw" back and forth between cold summers and moderate winters for an extended length of time. This is one of the biggest challenges confronting secular scientists who attempt to explain an ice age.

Over the years, secular scientists have come up with dozens of theories, and they are still proposing new ones.[2] Of course, the fact that scientists feel the need to come up with new theories is a clue that previous theories are not very convincing! Consider the following admissions by secular scientists during the late 1990s:

> Scientists have long known about the giant ice sheets that the Cenozoic ushered in. Even in the mid-19th century, they knew that glaciers had repeatedly raked swaths of Europe and North America in the not so distant past. Yet despite the efforts of marine geologists, atmospheric chemists, oceanographers, and more, no one knows what caused the ice ages. "We've been chewing on this problem for 30 or 40 years," says Alan Mix, an oceanographer at Oregon State University. "It's a killer." Adds Ralph Cicerone, dean of physical sciences at the University of California–Irvine, "It's embarrassing."[3]

Not much has changed since then. The precise causes for ice ages "remain controversial"[4] and are not "completely understood."[5]

However, many secular scientists believe astronomy is the key. Over tens of thousands of years, Earth would experience subtle changes in its motions as it orbits the sun. For instance, the current tilt of the Earth's axis is 23.4°, but over a period of about 41,000 years, this tilt would go from a minimum value of 22.1° to a maximum value of 24.5° and back again. Likewise, the earth's rotational axis wobbles like a spinning top, taking 26,000 years for the axis to trace out a complete circle.

The shape of the Earth's orbit becomes slightly more and less circular over an interval of about 100,000 years.[6] The top-like wobble combines with a slow rotation of the earth's orbit relative to the background stars to produce an overall cycle of 23,000 years. (A cycle of 19,000 years is also expected at times.) Because secular scientists believe the solar system is billions of years old, they think these motions have been going on for eons, and they feel free to extrapolate them backward into the "prehistoric" past.

These subtle changes cause small changes in the way sunlight falls on the earth, both with season and latitude. Many scientists believe the small changes control the timing of ice ages. Because this theory depends on the earth's astronomical motions, it is called the astronomical ice age theory—also known as the Milankovitch theory after the famous Serbian scientist who worked on it.[7]

In a nutshell, the theory states that there would be times in the distant past when more sunlight fell on the Northern Hemisphere high latitude ice sheets. At these times, the ice melted back, resulting in a warm interglacial. At other times, less sunlight fell on these ice sheets, allowing the ice sheets to grow in size, resulting in an ice age.

Biblical creationists disagree with this scenario. Although we agree with the physics and mathematics behind the calculations, we argue that the solar system is just 6,000 years old. Hence, it is invalid to extrapolate these present-day motions backward into a supposed prehistoric past.

Although the Milankovitch theory is popular, it has many problems that are well known among climate researchers.[8] The most obvious is that the changes in sunlight are very small, and it is extremely difficult to see how such small changes could themselves be responsible for an ice age. Famed evolutionist and astronomer Sir Fred Hoyle once ridiculed the theory:

> If I were to assert that a glacial condition could be induced in a room liberally supplied during winter with charged night-storage heaters simply by taking an ice cube into the room, the proposition would be no more unlikely than the Milankovitch theory.[9]

If the Milankovitch theory has such problems, then why is it so popular? There are a number of reasons for this. First, despite its problems, secular scientists have nothing better to replace it. So the Milankovitch theory has become the dominant theory almost by default. Second, many scientists seem fascinated with the idea that ice ages might occur at regular intervals, because this might allow them to predict the timing of future ice ages. Third, the theory became widely accepted because of an iconic paper published in the journal *Science* in 1976,[10] a paper that has since been shown to have serious miscalculations.[11]

In addition to being a weak explanation for an ice age, the Milankovitch theory also contributes to panic over climate change. Let's see how.

7

How Old-Earth Beliefs Fuel
Climate Alarmism

An evolutionary worldview definitely contributes to climate change alarmism. Not every scientist who accepts evolution and an old earth is worried about climate change, but nearly every scientist who *is* worried *does* believe those things. This includes scientists who are professing Christians. Texas Tech atmospheric scientist Dr. Katharine Hayhoe is an evangelical Christian, but she accepts the conventional story of Earth history involving millions of years. Like many, she is quite concerned about the issue.[1]

Remember that the heart of the global warming debate is the issue of climate sensitivity. A doubling of carbon dioxide would result in some warming, but the size of the warming depends on whether or not Earth's climate is stable. If the climate is unstable, factors called *positive feedbacks* will greatly amplify the warming, possibly resulting in a climate catastrophe. On the other hand, if the climate is stable, *negative feedbacks* will counter the positive feedbacks, preventing the earth from reaching an extreme climate state.

Remember also that one of the main problems with the Milankovitch theory is that the changes in sunlight thought to control the timing of ice ages are tiny. They are so small that it is difficult to see how they alone could be responsible for causing ice ages.

For this reason, secular scientists believe that *something*, some positive feedback, will greatly amplify these tiny changes in sunlight to bring about major climate change. They blame man-made carbon dioxide. Computer models can't simulate the sizes of supposed ice age temperature changes unless carbon dioxide is *assumed* to be a factor.

> Simulations with global climate models show that the amplitude [size] of glacial-interglacial temperature changes can only be reproduced if CO_2 changes are accounted for....This leads us to conclude that CO_2 changes are an important (feedback) factor in determining glacial-interglacial temperature changes although the ultimate cause of the ice ages are Earth's orbital cycles.[2]

Obviously, if carbon dioxide and other greenhouse gases really can greatly amplify the tiny changes in sunlight, then our climate must be unstable. Atmospheric scientist and creation researcher Dr. Larry Vardiman explains:

> A major result of this need for feedback mechanisms has been the development of a perspective that the earth's climate system is extremely sensitive to minor disturbances. A relatively minor perturbation could initiate a non-linear response which might lead to another "Ice Age" or "Greenhouse." Because of the fear that a small perturbation might lead to serious consequences, radical environmental policies on the release of smoke, chemicals, and other pollutants and the cutting of trees have been imposed by international agencies and some countries. If the basis for the Astronomical Theory is wrong, many of the more radical environmental efforts may be unjustified.[3]

Dr. Vardiman's reasoning is confirmed by climate alarmists themselves. One scientist explained how secular scientists "know" that Earth's climate system is extremely sensitive to small changes:

> The main limit on the [climate] sensitivity value is that it has to be consistent with paleoclimate data. A sensitivity which is too low will be inconsistent with past climate changes—basically if there is some large negative feedback which makes the sensitivity too low, it would have prevent-

ed the planet from transitioning from ice ages to interglacial periods, for example.[4]

The Milankovitch theory also contributes to concerns that sea levels could rise catastrophically. You may recall the 2004 movie *The Day After Tomorrow* that depicted global warming-induced catastrophic flooding of New York City. How does belief in this theory help stoke fears of such flooding?

Secular scientists think that data within seafloor sediments can be used to infer the amounts of ice on the planet at different times in the prehistoric past. Since the total mass of water on the earth is very nearly constant, global sea level *must* drop when more water is frozen during an ice age. Likewise, during warmer periods with less ice, global sea level *must* rise. So secular scientists believe that seafloor sediment data also tell them past global sea levels.

After secular scientists became convinced that the Milankovitch theory was correct, they began using it to assign ages to deep-sea sediments in a process called *orbital tuning*.[5] Secular scientists think they can use inferred past sea levels and the ages assigned to deep-sea sediments to calculate how fast sea levels rose in the prehistoric past. And guess what happens when secular scientists do this? They find that these calculated past rates have ominous implications for the future! Oceanographer Wolfgang Berger wrote:

> Just when can we expect to see a rapid rise of sea level, ten times higher than the present values of a few millimeters per year? We do not know. All we can say, *from experience with the many millennia of the ice-age records in the deep sea*, is that once melting starts, it stimulates further melting for centuries. Deglaciation keeps going once begun in earnest: a great example of the dilemma of the sorcerer's apprentice.[6]

What Berger calls "experience" is really just a Milankovitch interpretation of the seafloor sediment data, and creationists would argue that this interpretation is flat-out wrong.

Old-earth ideas also contribute to climate change alarmism by making false analogies between supposed past interglacials and the present climate. Remember that secular scientists think we are currently in the latest of many warm periods (interglacials) during the last 2.6 million years. Based on their interpretation of data from the Greenland ice cores, secular scientists concluded in 1993 that climate change could occur extremely rapidly.

> Climate in Greenland during the last interglacial period was characterized by a series of severe cold periods, which began extremely rapidly and lasted from decades to centuries. As the last interglacial seems to have been slightly warmer than the present one, its unstable climate raises questions about the effects of future global warming.[7]

Notice the logic here. Secular scientists were claiming that the previous warm period (interglacial) was characterized by severe climate change. Since we are supposedly in another interglacial, they argue by analogy that our present interglacial could *also* experience dramatic climate change.

Some creation scientists think that severe cold periods may have indeed occurred at the end of the Ice Age.[8] But they believe the Ice Age was caused by the Genesis Flood, a never-to-be-repeated event.[9] Because it was a unique time in Earth history, it is a mistake to use this unique time to draw conclusions about future climate change.

To be fair, the paper that made this analogy was published in 1993, and secular scientists may no longer agree with this particular interpretation of the data from Greenland. But I mention this example to show how secular scientists use an old-earth interpretation of climate data to draw conclusions about future climate change.

Of course, if the Milankovitch theory is wrong, then so are all these conclusions. The next chapter explores why there is no hard evidence that the theory is correct, even by secular, old-earth reckoning!

8

The Biggest Climate Scandal Yet?

As stated earlier, there are three main reasons secular scientists think Earth's climate is unstable. The first is the result of computer climate models, which we've seen have had a tendency to predict more warming than was later actually observed. The second is fraudulent, politicized research, which obviously should be given no weight at all in this controversy.

The third reason is related to the Milankovitch theory. If there's no supporting evidence for the theory, there is no reason for anyone, even evolutionists who believe in millions of years, to think the theory is correct. And without evidence for the Milankovitch theory, this last argument for an unstable climate collapses!

Most scientists believe the Milankovitch theory because of an iconic paper published in the journal *Science* in 1976.[1] Titled "Variations in the Earth's Orbit: Pacemaker of the Ice Ages," it seemed to provide support for the theory from deep seafloor sediments. This paper is so important in secular thinking that the prestigious journals *Nature* and *Science* both ran articles commemorating the paper's 40th anniversary in 2016.[2,3]

However, if one reads the "fine print" of that paper, the results that seemed to confirm the Milankovitch theory were critically dependent on an age assignment for an event that secular scientists themselves no longer accept as valid. This event was the most recent "flip" or reversal of the earth's magnetic field—during which the earth's north and south magnetic poles "traded" places—and in 1976 its assumed age was 700,000 years.[4] Of course, creation scientists think that age estimate is vastly inflated, and they would argue that these reversals in the earth's magnetic field were actually caused by the upheaval of the Genesis Flood.[5]

In any case, in the early 1990s secular scientists themselves revised this age assignment to 780,000 years.[6,7] An age change of 80,000 years might not sound like much (after all, secular scientists claim the earth is 4.6 *billion* years old), but that change was large enough to call into question the results of the Pacemaker paper.[8-10] Furthermore, secular scientists made other changes to the seafloor sediment data, changes that messed up the results even more![11,12]

Creation scientists pointed this out in 2016, but, to the best of my knowledge, secular scientists have yet to publicly acknowledge this very serious problem with the Pacemaker paper. In fact, you might have trouble believing that a mistake this big could go unnoticed for so long.

However, there is a very simple reason for this. The Pacemaker authors stated that they used the age for the most recent flip of the earth's magnetic field in their analysis, but they never stated *what that age was*. Instead, they referred back to a 1973 paper that explicitly gave the age as 700,000 years.[4] Most people, even most scientists, have not read that 1973 paper. For this reason, people tend to assume that the authors used an age of 780,000 years, since this is the currently accepted secular age assignment. In fact, even some books and scholarly articles incorrectly state that the Pacemaker authors used an age of 780,000 years.[13,14] But a careful reading of the relevant papers shows that this was not the case.

A paper by Dr. Maureen Raymo published in 1997 seems to provide some additional evidence for the Milankovitch theory.[15] But as we shall see a bit later, even this paper testifies to the weakness of the theory. If one reads between the lines, the second paper appears to be an attempt to discreetly prop up the Pacemaker paper!

Although secular scientists have never publicly acknowledged the problem with the Pacemaker paper, there are good reasons to think at least some of them are aware that evidence for the Milankovitch theory is very weak. Remember that secular scientists use this theory to assign ages to deep-sea sediments in a process called orbital tuning. They also use them to assign ages to the thick ice sheets of Greenland and Antarctica. If evidence for the Milankovitch theory is lacking, then the whole orbital tuning process is nothing but a giant exercise in circular reasoning.

Respected physicist Richard Muller of Berkeley made waves in 1996 when he and geophysicist Gordon MacDonald began pointing out still *another* problem with the Milankovitch theory that had been overlooked by researchers. Muller aired these criticisms at a 1996 conference. A 1997 *Science News* article said this about the meeting:

> Muller scored the most points at the meeting when he attacked a standard technique, called [orbital] tuning, that oceanographers used for dating layers in sediment cores. The task of dating these strata is difficult because sediments may accumulate more quickly during some eras and more slowly in others. To tell the age of layers between known benchmarks, researchers often use the Milankovitch orbital cycles to tune the sediment record: They assume that ice volume should vary with the orbital cycles, then line up the wiggles in the sediment record with ups and downs in the astronomical record.

> "This whole tuning procedure, which is used extensively, has elements of circular reasoning in it," says Muller. He argues that tuning can artificially make the sediment record support the Milankovitch theory.

> Muller's criticisms hit home with many researchers. "He scared the [expletive] out of them, and they deserved it," says [climate scientist W. S.] Broecker.[16]

So why were these researchers frightened? The answer is obvious. If hard evidence for the Milankovitch theory is lacking, then all their results from 20 years of orbital tuning are automatically suspect!

One scientist at the meeting was apparently concerned enough that she felt compelled to try to obtain additional evidence for the Milankovitch theory. *Science News* stated:

> Oceanographers soon rose to the challenge. In the August [1997] PALEOCEANOGRAPHY, Maureen E. Raymo of

the Massachusetts Institute of Technology presents an untuned sediment record that corroborates the ice age dates determined by tuning.[16]

Dr. Raymo is a highly regarded expert in the field of oceanography. Because of Muller's criticisms, she apparently felt the need to quickly come up with an alternate confirmation for the Milankovitch theory. But didn't the then 20-year-old Pacemaker paper *already* confirm the theory? Why did she feel an additional argument was needed?

Evidently, secular scientists realized, even back in 1996, that hard evidence for the Milankovitch theory was scanty at best. Hence Dr. Raymo's perceived need to find additional confirmation. It is not clear if secular scientists at this point were aware of the problem with the age of 780,000 years for the magnetic reversal, but one thing *is* clear: Dr. Raymo apparently felt uncomfortable hanging the entire argument for the Milankovitch theory on just the Pacemaker paper itself.

As we indicated in the earlier discussion of the Pacemaker paper, secular scientists have revised the age of the most recent flip of the earth's magnetic field. Now would be a good time to explain *why* they did that. Remember that in 1976 secular scientists claimed the age of this magnetic reversal was 700,000 years. By 1979 they had revised that age upward to 730,000 years.[17] This revision by itself was probably not enough to endanger the Pacemaker results. However, as noted above, in the early 1990s secular scientists advocated that the age of this reversal be revised upward *again* to 780,000 years. This required them to overrule their own age estimates based on radioactive dating.

But why? These scientists were attempting to use the Milankovitch theory and orbital tuning to assign ages to sediments in other locations. But they were having difficulty making the numbers work, so they arbitrarily raised the age of the reversal to *make them* fit!

Do you see why this is so outrageous? Secular scientists used an age of 700,000 years for this magnetic reversal to convince the world that the Milankovitch theory was correct.[1,4] Now they are using that same theory to argue that our climate is dangerously unstable and that we must take drastic action to save the planet. Yet, back in the early 1990s, they arbitrarily changed the age of that reversal because they were having trouble "fitting" data to the Milankovitch theory.[6,7] So instead of simply admitting that the theory couldn't handle all the data, they just changed some numbers to make it work. By doing so, they undermined their original argument for the Milankovitch theory, but today most people (including most scientists) are blissfully unaware of this.

Dr. Raymo's paper must have been an awfully big relief for the Milankovitch theory proponents who attended that 1996 meeting. It's lucky that Dr. Raymo was able to so quickly cobble together an alternate justification for the Milankovitch theory. Otherwise, 20 years of Milankovitch-based research might have gone up in smoke!

However, I don't find Dr. Raymo's 1997 paper to be terribly convincing. Her method implicitly repudiated the methodology used by the Pacemaker authors to get *their* age assignments. Furthermore, she did not use all the data then available to her, datasets that could have negatively impacted the results. Although she gave justifications for excluding those datasets, in my mind they sound more like excuses than reasons. And Muller and MacDonald have criticized her 1997 paper.[18]

I am not saying Dr. Raymo was being deliberately deceptive. However, because she believed the Milankovitch theory was correct, this may have (unconsciously, perhaps) influenced the choice of data she used in her analysis.

Here's what *should* have happened instead. As soon as secular scientists became aware that the Pacemaker results were invalid, whether in 1996 or 2016, the Pacemaker paper should have been retracted, as well as *all* the research based upon it. Of course, this would have invalidated decades of research, but that's the way the cookie crumbles! If secular scientists wanted to completely start over looking for evidence for the Milankovitch theory, fine. But they should have started over from scratch, rather than giving everyone the impression that the Pacemaker results were still valid. Because they did *not* do this, one can't help but suspect that they are simply unwilling to give up the theory and that they are *not* being objective in their analysis of the data.

So evidence for the Milankovitch (astronomical) ice age theory is weak at best and nonexistent at worst. Do you see why I titled this chapter "The Biggest Climate Scandal Yet?" It's even bigger than Michael Mann's infamous hockey-stick graph or the hacked East Anglia emails.

The Milankovitch theory is a major driver of global warming alarmism. It has led many scientists to conclude Earth has a dangerously unstable climate and that sea levels can rise catastrophically quickly. It also leads them to draw false analogies between our current warm climate and warm climates that supposedly occurred in the prehistoric past. Yet there is little, if any, evidence to back it up.

But if secular theories are so inadequate, then how do creationists explain the Ice Age?[19] In a nutshell, volcanic activity and rapid seafloor spreading during the Genesis Flood greatly warmed the world's oceans. These warmer oceans resulted in increased evaporation from the ocean surface, putting much more moisture into the air. This resulted in more rainfall, as well as greatly increased snowfall at mid and high latitudes and on mountains.

Sporadic post-Flood volcanic eruptions put tiny droplets and particles called *aerosols* into the atmosphere. These aerosols reflected sunlight,

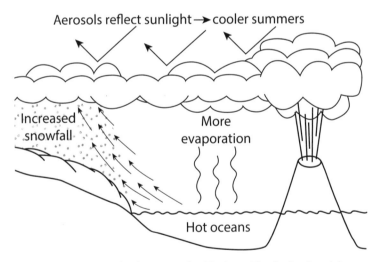

Figure 5. *Warm post-Flood oceans and residual post-Flood volcanic activity are the key to explaining the Ice Age.*
Image Credit: Susan Windsor.

resulting in colder summers. These cold summers prevented winter snow and ice from melting. As more snow and ice accumulated year after year, thick ice sheets formed (Figure 5).

However, this brief description can't really do justice to the creation Ice Age theory. It is truly impressive and explains many things that are still mysteries to secular scientists, such as how millions of wooly mammoths were able to thrive in Siberia during the Ice Age. We encourage interested readers to learn more at ICR.org/ice-age. The biblical model offers a much more satisfying explanation for the Ice Age than secular models, giving more reasons for Christians to stay levelheaded amid all the climate change "craziness."

Conclusion

In addition to scientific reasons, Christians have *theological* reasons to believe God created the earth to be stable. Despite the denials of secular scientists, there is overwhelming evidence that God created life on Earth and that this life did not originate through some evolutionary process. The evidence of design is so strong that even skeptics acknowledge that living things look designed for a purpose, even though they insist that this is merely an illusion. As Scripture states:

> For since the creation of the world His [God's] invisible attributes are clearly seen, being understood by the things that are made, even His eternal power and Godhead, so that they are without excuse. (Romans 1:20)

Scripture makes it clear that God created the entire universe, including the earth, moon, and sun (Genesis 1). Since an all-knowing, all-powerful, loving Creator designed our home, it is reasonable to think God designed its climate system to be stable, with mechanisms or negative feedbacks that prevent extreme climate states. This would definitely be true in God's original "very good" creation (Genesis 1:31), but we would expect this to be true even after Adam's Fall. God's promise to Noah after the Flood hints at a low climate sensitivity:

> And the LORD smelled a soothing aroma [of Noah's sacrifice]. Then the LORD said in His heart, "I will never again curse the ground for man's sake, although the imagination of man's heart is evil from his youth; nor will I again destroy every living thing as I have done. While the earth remains, seedtime and harvest, cold and heat, winter and summer,

and day and night shall not cease." (Genesis 8:21-22)

So even after the Flood, God promised a certain degree of reliability in Earth's climate system, which one would expect if the climate is stable.

There are good theological and scientific reasons to think Earth's climate is stable and global warming alarmism is unwarranted. Climate alarmism is distracting people—both Christians and non-Christians—from much weightier issues.

A supposed climate change emergency can sidetrack unbelievers from the true emergency that really *is* confronting them. Every single one of us will have to give an account to our Creator (Revelation 20:12). Because we have all sinned and violated His commandments, we can only tremble at this prospect. However, God has graciously provided a remedy. God became a man, the Lord Jesus Christ. He lived a perfect life and died as our substitute on the cross. He then rose from the dead three days later to demonstrate His victory over sin and death. All who are genuinely sorry for their sins and turn from them, believing that Jesus Christ is who He claimed to be, are promised salvation from God's righteous wrath.

Furthermore, the Lord Jesus Christ has tasked His Church with the proclamation of this saving gospel (Romans 1:16) in order to make disciples (Matthew 28:18-20). This earth, in its current form will not last forever (Revelation 21:1), no matter what we do or don't do, but resurrected men and women will live forever (Revelation 20:12–22:21).

Yes, God will hold us accountable for our stewardship of this planet, but He is also going to hold us accountable for what we did or did not do to fulfill His Great Commission. The Lord has given His Church a much more urgent task, and Christians should not allow themselves to be distracted from it by passing fads—including climate change alarmism.

Notes

Chapter 1: Don't Panic!

1. Witt, J. Doctor Doom, Eric Pianka, Receives Standing Ovation from Texas Academy of Science. Posted on evolutionnews.org April 3, 2006, accessed May 9, 2018.
2. Anderson, R. How Engineering the Human Body Could Fight Climate Change. The Atlantic. Posted on theatlantic.com March 12, 2012, accessed February 22, 2019.
3. Richardson, B. Pope Francis calls climate change a 'sin.' *The Washington Times*. Posted on washingtontimes.com September 1, 2016, accessed November 2, 2018; Climate Change: An Evangelical Call to Action. The Evangelical Climate Initiative. Posted on npr.org, accessed November 1, 2016.
4. McIntyre, S. and R. McKitrick. 2003. Corrections to the Mann. et al. 1998. Proxy Data Base and Northern Hemispheric Average Temperature Series. *Energy & Environment*. 14 (6): 751-771.
5. Biello, D. The Most Important Number in Climate Change: Just how sensitive is Earth's climate to increasing concentrations of carbon dioxide? *Scientific American*. Posted on scientificamerican.com November 30, 2015, accessed February 26, 2019.
6. Chandler, D. L. Explained: Climate sensitivity. MIT News. Posted on news.mit.edu March 19, 2010, accessed February 26, 2019.
7. Spencer, R. Global Warming 101. Posted on drroyspencer.com, accessed February 27, 2019.
8. Vaidyanathan, G. Did Global Warming Slow Down in the 2000s, or Not? *Scientific American*. Posted on scientificamerican.com February 25, 2016, accessed March 4, 2019.
9. Rogers, M. Global temperatures have dropped since 2016. Here's why that's normal. *Washington Post*. Posted on washingtonpost.com April 26, 2018, accessed March 4, 2019.
10. Rice, D. So you think you're cold? How does 88 below zero sound? *USA Today*. Posted on usatoday.com January 17, 2017, accessed March 1, 2019.
11. Blomberg, B. 2008. *Cool It: The Skeptical Environmentalist's Guide to Global Warming*. New York: Vintage Books, 13-21.

Chapter 2: Earth's Climate Is Stable

1. Biello, D. The Most Important Number in Climate Change: Just how sensitive is Earth's climate to increasing concentrations of carbon dioxide? *Scientific American*. Posted on scientificamerican.com November 30, 2015, accessed February 26, 2019.
2. Curry, J. A. and P. J. Webster. 1999. *Thermodynamics of Atmospheres and Oceans, Volume 65*. San Diego: Academic Press; Curry, J. 2002. *Encyclopedia of Atmospheric Sciences*. Holton, J. R., ed. 2002. Elsevier Science Ltd.
3. Judith Curry Curriculum Vitae. Posted on gatech.edu, accessed January 10, 2017.
4. Curry, J. A. Statement to the Subcommittee on Space, Science, and Competitiveness of the United States Senate: Hearing on "Data or Dogma? Promoting Open Inquiry in the Debate Over the Magnitude of Human Impact on Climate Change," accessed January 13, 2017; "Data or Dogma: Promoting Open Inquiry in the Debate over the Magnitude of Human Impact on Earth's Climate." *YouTube*. Video posted on youtube.com December 8, 2015, accessed January 10, 2017.
5. Curry, J. A. Statement to the Committee on Natural Resources of the United States House of Representatives: Hearing on "Climate Change: The Impacts and the Need to Act." Posted on curryja.files.wordpress.com, accessed February 27, 2019.
6. Lewis, N. and J. Curry. The impact of recent forcing and ocean heat uptake data on estimates of climate sensitivity. *Journal of Climate*. Published online April 23, 2018, accessed May 23, 2018.
7. Cook, J. How sensitive is our climate?: What is the possible range of climate sensitivity? Posted on skepticalscience.com September 8, 2010, accessed November 8, 2018.

8. Spencer, R. New Lewis & Curry Study Concludes Climate Sensitivity is Low. Posted on drroyspencer.com April 24, 2018, accessed October 17, 2018.
9. Thomas, B. Global Warming? Trees to the Rescue! *Creation Science Update*. Posted on ICR.org July 22, 2013, accessed February 28, 2019.
10. Lemonick, M. Freeman Dyson Takes on the Climate Establishment. *Yale360*. Posted on e360yale.edu June 4, 2009, accessed February 22, 2019.
11. Macfarlane, J. Amazing discovery of green algae which could save the world from global warming. *Daily Mail*. Posted on dailymail.co.uk January 4, 2009, accessed February 22, 2019.

Chapter 3: Could the Sun Be Affecting Weather and Climate?
1. Herschel, W. 1801. Observations tending to investigate the nature of the Sun, in order to find the causes or symptoms of its variable of light and heat; with remarks on the use that may possibly be drawn from solar observations. *Philosophical Transactions of the Royal Society of London*. 91: 265-318.
2. Vardiman, L. 2008. A New Theory of Climate Change. *Acts & Facts*. 37 (11): 10.
3. Hebert, L. III. 2011. Atmospheric Electricity Data from Mauna Loa Observatory: Additional Support for a Global Electric Circuit-Weather Connection? Ph.D. dissertation, University of Texas at Dallas.
4. Hebert, L. III, B. A. Tinsley, and L. Zhou. 2012. Global Electric Circuit Modulation of Winter Cyclone Vorticity in the Northern High Latitudes. *Advances in Space Research*. 50 (6): 806-818.
5. Hebert, L. III, B. A. Tinsley, and L. Zhou. 2011. Is the Global Electric Circuit Modulating Winter Cyclone Vorticity in the Northern High Latitudes? American Geophysical Union poster abstract. Archived at adsabs.harvard.edu/abs/2011AGUFMAE31A0255H, accessed November 30, 2018.

Chapter 4: Faulty Computer Models
1. Spencer, R. New Lewis & Curry Study Concludes Climate Sensitivity is Low. Posted on drroyspencer.com April 24, 2018, accessed October 17, 2018. Emphases and ellipses in original.
2. Stocker, T. F. et al, eds. 2013. *Climate Change 2013: The Physical Science Basis*. New York: Cambridge University Press.
3. Dyson, F. Heretical Thoughts about Science and Society. *Edge*. Posted on edge.org August 7, 2007, accessed February 22, 2019.
4. Curry, J. JC in transition. Posted on judithcurry.com January 3, 2017, accessed January 10, 2017.
5. Gayathri, A. Nobel Laureate Ivar Giaever Quits Physics Group over Stand on Global Warming. *International Business Times*. Posted on ibtimes.com September 15, 2011, accessed January 10, 2017.
6. Revkin, A. C. A Physicist's Climate Complaints. *New York Times*. Posted on nytimes.com October 15, 2010, accessed January 10, 2017.

Chapter 5: Politics and Junk Science
1. Hutson, W. T. Top Meteorologist: Climate Change 'Nothing but a lie.' *Breitbart*. Posted on breitbart.com October 23, 2014, accessed February 22, 2019.
2. Stocker, T. F. et al, eds. 2013. Summary for Policy Makers. In *Climate Change 2013: The Physical Science Basis*. New York: Cambridge University Press.
3. Idso, C., R. M. Carter, S. F. Singer, eds. 2013. Summary for Policy Makers. In *Climate Change Reconsidered II: Physical Science. 2013 Report of the Nongovernmental International Panel on Climate Change (NIPCC)*. Chicago: The Heartland Institute.
4. McIntyre, S. and R. McKitrick. 2005. Hockey sticks, principal components, and spurious significance. *Geophysical Research Letters*. 32 (3): L03710; McKitrick, R. What is the 'Hockey

Stick' Debate About? Invited Special Presentation to the Conference Managing Climate Change—Practicalities and Realities in a Post-Kyoto Future before the Parliament House, Canberra, Australia, April 19, 2005.

5. Muller, R. Global Warming Bombshell. *MIT Technology Review*. Posted on technologyreview. com October 15, 2004, accessed November 7, 2018.

6. *"A Disgrace to the Profession": The World's Scientists – In Their Own Words – on Michael E Mann, His Hockey Stick, and Their Damage to Science*, vol. 1. M. Steyn, ed. Woodsville, New Hampshire: Stockade Books.

7. Thomas, B. Leaked Emails May Show Global Warming Research Is a Fraud. *Creation Science Update*. Posted on ICR.org December 3, 2009, accessed February 28, 2019.

8. Jacoby, J. Climategate: Dissent on ice. Posted on boston.com December 2, 2009, accessed February 28, 2019.

9. Financial Statements for the Year to 31 July 2017 (PDF). University of East Anglia. page 18. Retrieved December 13, 2017.

10. Hebert, J. Prosecute Climate-Change Skeptics? *Creation Science Update*. Posted on ICR.org October 12, 2015, accessed February 22, 2019.

11. Curry, J. RICO! Posted on judithcurry.com September 17, 2015, accessed February 22, 2019.

12. Vaughan, A. World court should rule on climate science to quash sceptics, says Philippe Sands. Posted on theguardian.com September 18, 2015, accessed February 22, 2019.

Chapter 6: Secular Climate Change Theories

1. Kerr, R. 1997. Upstart Ice Age Theory Gets Attentive But Chilly Hearing. *Science*. 277 (5323): 183-184.

2. University of Royal Holloway London. New theory on cause of ice age 2.6 million years ago. ScienceDaily. Posted on sciencedaily.com June 27, 2014, accessed November 2, 2018.

3. Watson, T. 1997. What causes ice ages? *U.S. News & World Report*. 123 (7): 58-60.

4. Ice age. *New World Encyclopedia*. Posted on newworldencyclopedia.org, accessed December 12, 2016.

5. Ice Ages. BBC. Posted on bbc.co.uk, accessed December 12, 2016.

6. Actually, this is an oversimplification. More careful astronomical calculations show two closely spaced cycles, one of 95,000 years, and another of 125,000 years. Muller, R. A. and G. J. MacDonald. 2000. *Ice Ages and Astronomical Causes: Data, Spectral Analysis and Mechanisms*. Chichester, UK: Praxis Publishing, 13, 40-45. In fact, this particular criticism of the theory is alluded to in Chapter 7.

7. Milanković, M. 1941. *Canon of insolation and the Ice-Age problem*. Belgrade, Serbia: Special Publication of the Royal Serbian Academy, vol. 132.

8. Cronin, T. M. 2010. *Paleoclimates: Understanding Climate Change Past and Present*. New York: Columbia University Press, 130-139.

9. Hoyle, F. 1981. *Ice, the Ultimate Human Catastrophe*. New York: Continuum, 77.

10. Hays, J. D., J. Imbrie, and N. J. Shackleton. 1976. Variations in the Earth's Orbit: Pacemaker of the Ice Ages. *Science*. 194 (4270): 1121-1132.

11. Hebert, J. New Calculations Melt Old Ice Age Theory. *Creation Science Update*. Posted on ICR.org September 14, 2016.

Chapter 7: How Old-Earth Beliefs Fuel Climate Alarmism

1. Cahalane, L. *Decoding the Weather Machine*. NOVA. Aired April 18, 2018.

2. Schmittner, A. Paleoclimate. Introduction to Climate Science. Online text for Oregon State University, College of Earth, Ocean, and Atmospheric Sciences. Posted on library.open. oregonstate.edu, accessed January 4, 2019.

3. Vardiman, L. 2001. *Climates Before and After the Genesis Flood*. El Cajon, CA: Institute for Creation Research, 79.

4. Nuccitelli, D. How sensitive is our climate? Skeptical Science. Posted on skepticalsicence. com, accessed November 13, 2018.

5. Hebert, J. 2016. Deep Core Dating and Circular Reasoning. *Acts & Facts*. 45 (3): 9-11.

6. Berger, W. H. 2012. Milankovitch Theory—Hits and Misses. Scripps Institution of Oceanography, University of California San Diego, La Jolla, California, 16. Emphasis added.

7. Greenland Ice-Core Project (GRIP) Members. 1993. Climate Instability During the Last Interglacial Period Recorded in the GRIP Ice Core. *Nature*. 364: 203-207.

8. Oard, M. J. 2005. *The Frozen Record*. El Cajon, CA: Institute for Creation Research, 125-128.

9. Hebert, J. 2018. The Bible Best Explains the Ice Age. *Acts & Facts*. 47 (11): 10-13.

Chapter 8: The Biggest Climate Scandal Yet?

1. Hays, J. D., J. Imbrie, and N. J. Shackleton. 1976. Variations in the Earth's Orbit: Pacemaker of the Ice Ages. *Science*. 194 (4270): 1121-1132.

2. Hodell, D. A. 2016. The smoking gun of the ice ages. *Science*. 354 (6317): 1235-1236.

3. Maslin, M. 2016. Forty years of linking orbits to ice ages. *Nature*. 540 (7632): 208-210.

4. Shackleton, N. J. and N. D. Opdyke. 1973. Oxygen Isotope and Palaeomagnetic Stratigraphy of Equatorial Pacific Core V28-238: Oxygen Isotope Temperatures and Ice Volumes on a 10^5 and 10^6 Year Scale. *Quaternary Research*. 3: 39-55.

5. Humphreys, D. R. 1986. Reversals of the Earth's Magnetic Field During the Genesis Flood. In *Proceedings of the First International Conference on Creationism*. R. E. Walsh, C. L. Brooks, and R. S. Crowell, eds. Pittsburgh, PA: Creation Science Fellowship, 113-123

6. Shackleton, N. J., A. Berger, and W. R. Peltier. 1990. An Alternative Astronomical Calibration of the Lower Pleistocene Timescale Based on ODP Site 677. *Transactions of the Royal Society of Edinburgh: Earth Sciences* 81 (4): 251-261.

7. Hilgen, F. J. 1991. Astronomical Calibration of Gauss to Matuyama Sapropels in the Mediterranean and Implication for the Geomagnetic Polarity Time Scale. *Earth and Planetary Science Letters* 104 (2-4): 226-244.

8. Hebert, J. 2016. Milankovitch Meltdown: Toppling an Iconic Old-Earth Argument, Part 1. *Acts & Facts*. 45 (11): 10-13.

9. Hebert, J. 2016. Milankovitch Meltdown: Toppling an Iconic Old-Earth Argument, Part 2. *Acts & Facts*. 45 (12): 10-13.

10. Hebert, J. 2017. Milankovitch Meltdown: Toppling an Iconic Old-Earth Argument, Part 3. *Acts & Facts*. 46 (1): 10-13.

11. Hebert, J. More Problems with Iconic Milankovitch Paper. *Creation Science Update*. Posted on ICR.org July 13, 2018, accessed March 1, 2019.

12. Hebert, J. 2017. The "Pacemaker of the Ice Ages" Paper Revisited: Closing a Loophole in the Refutation of a Key Argument for Milankovitch Climate Forcing. *Creation Research Society Quarterly*. 54: 133-148.

13. Woodward, J. 2014. *The Ice Age: A Very Short Introduction*. Oxford, UK: Oxford University Press, 97.

14. Nisancioglu, K. H. 2010. Plio-Pleistocene Glacial Cycles and Milankovitch Variability. In *Climates and Oceans*. J. H. Steele, ed. Amsterdam, The Netherlands: Academic Press, 344-353.

15. Raymo, M. E. 1997. The timing of major climate terminations. *Paleoceanography and Paleoclimatology*. 12 (4): 577-585.

16. Monastersky, R. 1997. The Big Chill: Dust dust drive Earth's ice ages? *Science News*. 152 (14): 220.

17. Muller, R. A. and G. J. MacDonald. 2000. *Ice Ages and Astronomical Causes: Data, Spectral Analysis, and Mechanisms*. Chichester, UK: Praxis Publishing, 149-160.

18. Ibid, 205-206.

19. Hebert, J. 2018. The Bible Best Explains the Ice Age. *Acts & Facts*. 47 (11): 10-13.